Ho
LIVE ON
MARS

How to...

LIVE ON
MARS

By CLIVE GIFFORD

Illustrated by
Scoular Anderson

ROVER 1

OXFORD
UNIVERSITY PRESS

OXFORD
UNIVERSITY PRESS

Great Clarendon Street, Oxford OX2 6DP

Oxford University Press is a department of the University of Oxford.
It furthers the University's objective of excellence in research, scholarship,
and education by publishing worldwide in

Oxford New York

Athens Auckland Bangkok Bogotá Buenos Aires Calcutta
Cape Town Chennai Dar es Salaam Delhi Florence Hong Kong Istanbul
Karachi Kuala Lumpur Madrid Melbourne Mexico City Mumbai
Nairobi Paris São Paulo Singapore Taipei Tokyo Toronto Warsaw

with associated companies in Berlin Ibadan

Oxford is a registered trade mark of Oxford University Press
in the UK and in certain other countries

Series devised by Hazel Richardson

Text copyright © Clive Gifford 2000

British Library Cataloguing in Publication Data available

ISBN 0-19-910738-6

1 3 5 7 9 10 8 6 4 2

Printed in the United Kingdom

Contents

HOW TO LIVE ON MARS

Why go to Mars? Well, the first answer is because it's there. And the second is because – just maybe – we can. We've been up into space. We've landed on the Moon. The next step? Mars. We've already sent unmanned probes to check out what it's like.

But it's been years since anyone went to the Moon – why Mars now? What's changed? Two things. One is our technological know-how. Computers today are thousands of times more powerful, and a great deal cleverer, than those used on the early Apollo missions to the Moon. The other big change is what we know about space. In 1998, the Lunar Prospector mission saw signs that may possibly lead to the discovery of ice on the Moon. This opens up the real possibility of using the Moon as a base for manned missions further afield – in particular to Mars.

If you're planning a trip to Mars, you might want to ask some questions:

- ○ Is there life on Mars?

- ○ Was there life there in the past?

- ○ Can we create a human colony on its surface?

- ○ Will there be a space bus to Mars in the near future?

- ○ If there is a space bus, how do you go about booking a seat on it?

Well, that's what this book is all about.

Please fasten your seat belts.
Our destination is Mars colony ZXC 0297.
Expected flying time, 6 months.

WHY LEAVE EARTH?

Human beings are inquisitive. There are always people who want to know what's around the next corner, beyond the horizon, or across the sea. Explorers have travelled to the far reaches of Earth. Mountaineers have climbed the highest mountains and divers have made it to the deepest parts of the ocean. So what's left to explore? Space, that's what!

After hundreds of years of studying the night sky with telescopes, people became very excited when they were able to send first machines and then people into space.

Yippee!

In the 1950s we put the first artificial satellites in space. Within a few short years, astronauts were sitting aboard cramped spacecraft that orbited the Earth. In 1969 the first person landed on the Moon. All this happened very fast, but the next steps – establishing a base on the Moon or sending manned missions to Mars – have taken an age. Why? Well there are two reasons. Scientists are still working on how to send people as far as Mars. And a manned mission to Mars would be very expensive.

Some people believe that sending people and machines into space was, and continues to be, a complete waste of time. We didn't actually find much on the Moon, they say, and it's never going to be possible for lots of people to live comfortably there, or on any of the planets.

It's certainly a very expensive hobby for governments who have plenty of other things on Earth to spend their money on.

But our travels into space have taught us a great deal about the Earth, the Universe, our Moon, the planets around us, life, and ourselves. It's knowledge that we couldn't have learned in any other way. And there's more ...

Spacey spin-offs ....

Critics of space exploration say that all we've achieved from the space missions are anti-gravity pens that can write upside-down and thermal blankets. But there have been lots of spin-off benefits from the space programme. Direct advances in medicine, robotics and new materials have come from research into space. Indirect advances have come from the needs of the space programme driving other industries to improve. The best example of this is the computer industry. When the space programme took off in the 1960s, computers were the size of a large room and very slow. The need for small and powerful computing on spacecraft helped to lead at the end of the 1960s to the development of the silicon chip. This is the tiny integrated circuit that helps computers to be what they are now – small, slim and super-powerful.

But the biggest benefit from space exploration has been satellites. These guys orbit the Earth thousands of kilometres above our heads and perform many incredibly useful jobs. Weather satellites help us to predict weather patterns, and warn of storms and hurricanes. Communications satellites beam live TV pictures and thousands of phone calls all around the world.

And who knows what benefits the new International Space Station will bring once it starts working in a couple of years' time? Better treatments and cures for certain diseases? Brand new materials? We'll have to wait and see. Curious? Of course you are!

Your curiosity is similar to that of inquisitive scientists the world over. Curiosity is the main reason why the space station is being built.

It's also the reason why there are more and more missions to Mars. But why Mars? Well, we've sent machines and people to the Moon: Mars is the obvious next step.

HI, I'M MARS

Mars is the fourth furthest planet from the Sun, and one of Earth's next-door neighbours (the other is Venus). But sometimes it takes a while to get to know your neighbours.

This is what we've found out about Mars so far – but we'd really like to call in for a closer look.

Some Martian facts:

○ Measuring 6792 kilometres in diameter, Mars is about half the size of Earth.

○ A day on Mars is just half an hour longer than a day on Earth. But the planet takes

13

I'm only ten Mars years old!

WELCOME TO MARS

almost twice as long as Earth to orbit the Sun – 687 Earth days equal one Mars year.

○ Mars is more like Earth than any other planet in the Solar System. This is one of the main reasons why we're all so interested in it. Compared to the other planets in our Solar System, it's remarkably friendly, as the following fact file shows.

PLANETARY FACT FILE

Mercury: Forget it, far too close to the Sun!

Venus: Heavily clouded, but still so hot that it would melt lead. Also, the atmosphere is about 90 times as thick as Earth's and is made mainly from choking carbon dioxide.

Jupiter: Biggest planet in the Solar System, but no solid surface and violent storms would wreck any man-made structures.

Saturn, Uranus, and Neptune: These planets lack a solid surface to land on. Also, they are a very long way away, and are surrounded by clouds.

Pluto: Far too far away, and not very interesting anyway.

Saturn

Uranus

Neptune

Compared to icy-cold Neptune and Uranus, boiling-hot Mercury, and just plain scary Venus, Mars looks like a holiday camp. But let's have a closer delve into the surface and see what we find.

Desert ... Rats! ....

Early astronomers were very excited when they looked at Mars's surface through telescopes and saw large red-orange areas. They thought these were sandy deserts, and where there's a desert there's bound to be an oasis with trees and water, right?

Wrong!

So far, no telescope, however powerful, has found a trace of a palm tree. The real news is that the surface may look like desert, but it's ice-cold. The average temperature on Mars is –23 °C.

That single temperature doesn't tell the whole story. Mars's temperature varies greatly throughout the year.

At its warmest, the area around Mars's equator can reach a sultry 27 °C; while the lowest recorded temperature anywhere on the planet has been –128 °C. In a single day, Mars can have temperature differences of up to 100 °C.

Mars's surface does look a bit like a desert, though. Loose stones, soil and sand cover a lot of it. The soil on Mars is made up mainly of silicon and oxygen, mixed with metals like magnesium and iron. When the iron and oxygen react together they form iron oxide – or rust. This is what gives Mars its distinctive red colour.

Move to Mars—
MAKE YOUR OWN MARTIAN SOIL

WHAT YOU'LL NEED
* steel wool
* a saucer
* rubber gloves
* a paper towel

WHAT TO DO

1 Fold the paper towel in half twice and place it in the saucer.

2 Run warm water over the steel wool for a while. Then place the wet steel pad on the tissue, and find a place for the saucer that won't be disturbed for a week or so.

3 After two weeks, put on the rubber gloves and rub the pad between your fingers.

WHAT HAPPENS?
The pad should disintegrate into a reddish powder. This is rust or, to give it its scientific chemical name, iron oxide.

Atmosphere......................................

The atmosphere of Mars is made up of carbon dioxide, nitrogen, argon, oxygen, water vapour and carbon monoxide. Water is vital to life – so scientists became very interested in precisely how much water there is in the Martian atmosphere. The answer is, not very much.

If you could take the entire atmosphere of Mars and wring it out, the water would fill a few dozen swimming pools.

Is this the best we can do?

Now that may sound a lot, but compared with the masses of water present in the Earth's atmosphere, it's almost nothing.

Missions to Mars ....

There have already been well over twenty Mars missions. Unfortunately, many have been complete disasters. The Soviet Union launched eight missions to Mars in the 1960s: some were landers and others were fly-by missions. Not one of them made it.

Oh no, missed again!

The first hard facts about the atmosphere and surface of Mars came with the *Mariner 4* spacecraft, launched in 1964. This was the first successful fly-by of Mars. *Mariner 4* travelled to within 10,000 kilometres of the planet's surface, found that carbon dioxide was the main gas in the atmosphere, and took 22 photographs. Mind you, it did take *Mariner 4* eight hours to send each fuzzy picture back to Earth. Still, scientists were extremely excited. They had in their hands the first detailed pictures of Mars's surface.

Whipping up a storm ....

Having no rain or surface water makes most of Mars's surface as dry as a bone. Storms on Mars really whip up the soil and sand, and create enormous red dust clouds. These can cover large parts of the planet, sometimes even all of it, for months at a time. Early astronomers saw dark patches on Mars and hoped these were lush forests. But sadly not. The darker patches are areas where the winds have blown away the loose stuff on the surface, to reveal the solid rock beneath.

Move to Mars–
SEE HOW WINDS SHIFT DRY MARTIAN SOIL

You can see in the following experiment how winds move around the loose, dry material on the surface of Mars.

WHAT YOU'LL NEED

* a pan or deep tray
* a hole punch
* a piece of paper
* a plastic plant-spray bottle

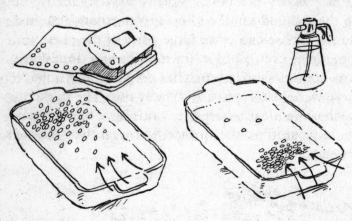

WHAT TO DO

Punch about 50 holes in the paper and empty the small paper circles from the hole punch into the pan. Blow across the paper circles. Now spray the paper circles with water from the plant spray until they are damp.

WHAT HAPPENS?

The dry paper circles move easily when you blow them. When they are damp the circles are heavier, and they stick together, so they are harder to move.

Caps off..

Mars has two polar ice-caps that seem to grow in winter and shrink in summer. The ice-cap at the north pole is bigger than the one at the south pole.

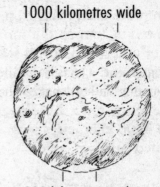

1000 kilometres wide

300 kilometres wide

The ice may not necessarily be frozen water. Scientists hope it is, because if so, future missions may be able to melt part of the caps to provide water. However, until a mission successfully examines the area and lets us know, we can't be sure. Many scientists think the caps are made at least partly of frozen carbon dioxide. You may have seen a form of frozen carbon dioxide in action. It's called dry ice and often provides the 'smoke' in films and at pop concerts.

THE MONSTER FROM MARS

Valley deep, mountain high.....................

A number of unmanned missions to Mars have allowed us to map the surface in some detail. Two of Mars's most prominent features are a giant extinct volcano and a massive valley. The Valles Marineris (Mariner Valley) makes the Grand Canyon look like a hole in the road. It's a staggering 600 kilometres wide and 4500 kilometres long. That's as long as the United States is wide.

Olympus Mons – Mars's highest mountain – rises 23 kilometres above ground level. That's a few metres short of three times the height of Mount Everest. Its crater at the top would swallow Birmingham or Manchester with no problem.

Olympus Mons and the Valles Marineris were photographed for the first time by the *Mariner 9* spacecraft. *Mariner 8* had been a complete failure and plopped into the sea just off the island of Puerto Rico. But – wow! – was *Mariner 9* a success! It took 7329 photographs of parts of the Mars surface.

No pressure

The atmosphere on Earth is really very heavy. Every cubic metre of air weighs more than a kilogram. And that's the same as having a whole kilo pressing down on every centimetre of your body all the time!

Move to Mars—
SEE ATMOSPHERIC PRESSURE IN ACTION

WHAT YOU'LL NEED
* a flexible plastic bottle
* a good breath

WHAT TO DO
Breathe out completely, then put your lips firmly round the bottle opening and breathe in deeply. You're sucking the air out of the bottle.

WHAT HAPPENS?

The bottle buckles as its sides go in. What is forcing the sides in? Air pressure, that's what. You're sucking air and air pressure out of the bottle, causing the air pressure outside the bottle to become greater than that inside the bottle. That's what forces the sides in.

But the atmosphere on Mars is incredibly thin and exerts much less pressure on the planet's surface – less than one hundredth of the pressure here on Earth.

A matter of some gravity ....

Mars has some gravity, but not as much as Earth.
The gravity on Mars is around a third of what we
experience on Earth. This will make every astronaut
on Mars a better golfer – with golf balls travelling a
lot, lot further. Much more importantly, it may allow
lightweight, inflatable buildings and structures to be
built easily.

> Er ... did we
> remember to fasten
> down the inflatable
> building?

But we're getting ahead of ourselves. Before we land
on Mars and start building, let's have a look at how
other people have seen Mars throughout history.

CHAPTER THREE

THE LURE OF THE RED PLANET

Compared with the other objects in the sky, Mars has always stood out as a bit unusual. It glows red, while most of the stars are white. And it has this strange looping movement across the night sky. The ancient Egyptians called Mars *seked-ef em khetkhet* – one who travels backwards. Mars *does* appear to travel backwards for a short time each year, but today we know that that isn't really what happens. Because Earth is closer to the Sun than Mars, its orbit isn't so long. So when the Earth 'overtakes' Mars as it orbits the Sun, it looks as if Mars is going backwards.

Fortunately for us, our glowing red star later got a much shorter name, and one that's much easier to spell. The red planet was named after the god of war – Ares to the Ancient Greeks, but Mars to the Romans. Many Romans worshipped the god of Mars as their protector. Some thought that the planet's red colour came from running blood – scary!

New ideas ....

Nothing much more happened about investigating Mars until the 16th and 17th centuries. By this time, the printing press had been invented and lots more people were reading books. This triggered off an explosion in thought and learning, much of it about the stars and planets in the sky. People read the works of the Ancient Greeks and Romans and discussed new ideas. Mind you, if your ideas threatened the churches or rulers of where you lived, you were in serious trouble.

For instance, at that time most people thought that our little Earth was at the centre of the Universe, and that the Sun, Moon, stars, planets, and everything circled round it. A few people suggested that maybe we'd got it wrong. Maybe the Sun was the main man – and the Earth, and Mars, and all the other planets just circled round him. They were right, of course – but hardly anyone wanted to know, especially not the Church. And in those days the Church could take all sorts of measures to stop you talking.

The Duelling Dane

Denmark, 1576

The first great Mars explorer was a Danish nobleman called Tycho Brahe. He was completely devoted to astronomy but still found enough time to lead a colourful life. He lost much of his nose in a duelling accident and fitted himself with a new one he made out of gold, silver and wax.

Old Metal Nose was a remarkable astronomer. In 1576, he set up an observatory on an island near Copenhagen. For the next twenty years, he observed and kept incredibly detailed records of over 700 stars. His calculations of the positions of planets, especially Mars, were incredibly accurate. All the more amazing was that Brahe did all this before the invention of the telescope. What did he use, then? His eyes, that's all.

Tycho took on a student called Johannes Kepler. They didn't get on too well. As a result, Brahe gave Kepler some really tough homework — studying the movement of the Solar System's most erratic planet. You guessed it, Mars.

After Brahe died in 1601, Kepler discovered that Mars travels around the Sun in an oval-shaped orbit called an ellipse. He also calculated that Mars might have two moons, but never got to see them. Poor Kepler wrote to the Mr Telescope of his time, Italian scientist, Galileo Galilei. He begged and pleaded for a loan of one of Galileo's 'seeing machines', but the Italian refused. What a meanie!

Handy Hans
Holland, 1608

The telescope had been invented in 1608 by a Dutch spectacle maker called Hans Lippershey. Fixing lenses in a row inside a tube, he found that he could make things appear several times bigger than they really were.

Great Galileo

Italy, 1609

But it was Galileo who really developed the telescope. Just a year after Lippershey's invention, Galileo built a telescope that could magnify 30 times. With the help of his telescopes, Galileo made many important discoveries about our Solar System.

What I discovered today:
1 There are four moons around Jupiter.
2 The Earth's moon is not smooth but full of mountains and craters.
3 The Milky Way is made up of lots of individual stars.
4 Venus has phases like the Moon.

Telescopes gradually got bigger and better. Much thanks is due to the Dutch scientist, Christiaan Huygens. He found ways of grinding and polishing lenses that made them work better. He used his new, improved telescope to good effect. He discovered a moon around Saturn and was the first person to produce a sketch of Mars that had real features on it.

Moons of Mars ....

Remember that prediction of Kepler's? You know, the one where he thought there were two moons orbiting Mars. Well, it took over 200 years to prove him right. In 1877, an American astronomer called Asaph Hall discovered not one, but both moons. He was given the honour of naming them and looked back into Greek myths. The Greek god of war (from whom Mars got its name) was always joined in battle by his two children, Deimos and Phobos. And that's what Asaph Hall named the two moons.

Don't mess with me and Phobos!

Canal fever ....

1877 was a big year for Mars watchers. Apart from the moons, there was an announcement from the Italian astronomer, Giovanni Virginio Schiaparelli. He noticed straight lines cut into the surface of Mars, which was all very well. But what followed was a bit of a mistake.

Words often have many meanings and can get translated wrongly. In Italian, Schiaparelli had written 'channels', but in English it was translated as 'canals'. The Suez canal was being built at the time, and people were canal-crazy. To them, canals meant creature-made waterways – which meant there was life on Mars!

We now know that the 'canals' which caused so much fuss were purely imaginary, or at best optical illusions. There are channels cut into the Martian surface, but when the astronomer Patrick Moore compared Schiaparelli's canal maps with the maps of Mars we have today they didn't fit at all. Scientists think that these channels were probably formed by water and ice eroding the rock. This means that Mars was once a planet on which water flowed.

Move to Mars–
SEE HOW ICE HAS FORCE

WHAT YOU'LL NEED
* a plastic drinking straw
* some modelling clay
* a glass of water

WHAT TO DO

1   Suck some water up into the straw. Put your finger over the end of the straw, so that the water stays in while you plug the open end with some of the modelling clay.

2   Turn the straw upside-down and plug the other end with the rest of the modelling clay.

3   Put the straw in the freezer and wait for three hours. Open the freezer and look at the straw.

WHAT HAPPENS?

The water has turned to ice and one of the clay plugs has been forced out. When water turns to ice it expands and pushes with force. When water gets into cracks in rock and freezes, it can push hard enough to crack some rock.

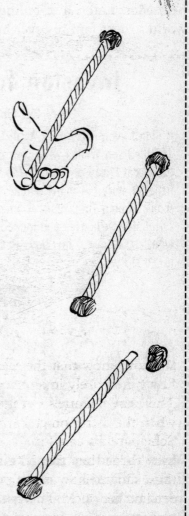

Astronomers, scientists and the general public continued getting their knickers in a twist about civilizations on Mars. Martians became a popular theme for books, comics, cartoons and films. H. G. Wells' book *War Of The Worlds*, which told of Martians landing on Earth and taking over, was a bestseller. Later, a re-telling of the story became the world's most famous ever hoax.

Invasion from Mars!

USA, 30 October 1938

It didn't even occur on 1 April, but on the eve of Halloween in 1938. Orson Welles, who was 23 at the time, staged a radio broadcast based on *War Of The Worlds*. It was announced as a fictional story but many people still believed that Martians had really landed. Thousands of Americans were panic-stricken. Many fortified their houses and loaded their shotguns waiting for the invasion. It took a few days to convince everyone that it was just a story.

The Martians have landed!

Welles' broadcast turned him into an overnight star. It also showed how many people really believed there might be life on Mars. Want to know more about that? Read on.

CHAPTER FOUR

ALIEN LIFE

I can assure you there is absolutely no chance of there being life in our Solar System, or in any other nearby solar system.

People all over the world are completely fascinated by the idea that alien life might exist. That's why TV programmes, films and books about aliens are such a success. It's also why books on unidentified flying objects (UFOs) sell like hot cakes.

Yuk! Horrible aliens!

Did they land?

The first attempt to get a machine on to Mars to find out whether little green men and women live there happened in 1969. The Soviet Union sent seven missions to land on Mars. Scientists believe that some of these actually made it to the surface, but contact was lost with all seven of them. Perhaps, in the future, we'll find out what really happened.

The HoaX Files ....

To be frank, most of the so-called 'aliens have visited Earth in the past' theorists have been unmasked as hoaxers. And while we're spoiling the party, we might as well go further. The vast majority of UFO sightings are genuinely things like aircraft, weather balloons, optical illusions and natural phenomena like St Elmo's fire – a type of electrical charge that creates a glow around tall objects. The planet Venus is always being reported as a UFO. But people want to believe life from other planets and stars exists, so they try really, really, really hard to make it happen.

Move to Mars–
FAKE YOUR OWN UFO PICTURE

Let's see how easy it is to make your own fake UFO photograph.

WHAT YOU'LL NEED
* a camera with film
* a friend
* a round metal baking dish

WHAT TO DO
You and your friend need to be outdoors in plenty of space. ➤

Ask your friend to have the camera at the ready. It should be focused on the background. Gently toss the baking dish as you would a frisbee. Try to throw it at an angle so that the bottom of the dish partly faces the camera — this will give it a more convincing shape in the photo. Take several photos and take the film to the processors.

WHAT HAPPENS?
When the pictures come back, one, maybe more, should be half-convincing. The object in the sky will be blurred, giving the impression that it is moving fast.

Party-poopers

Scientists aren't heartless monsters with no emotions or passions. On the contrary, many of them went into science because they hoped they could prove something that other people just sniggered at. There's nothing more that most space scientists would like than for alien life to show up tomorrow. But scientists also have a duty to prove things scientifically whenever they can. Even though most sightings and theories of alien life have proved to be nonsense, there do remain a few unsolved mysteries ...

How come ancient peoples built really clever star observatories and calendars. Surely they had alien help?

What about the many ancient paintings of creatures with what look like space suits on?

As we said, no one truly knows and perhaps never will. Most people think the lines in the sand and the space suits just came from the imaginations of ancient artists. And there's a good chance that the carvings of the movements of the Sun and stars were the work of dedicated astronomers from ancient cultures. You have to admit that these explanations sound more likely than an alien race being able to travel many hundreds of millions of kilometres while still needing 21st-century airport runways to land.

The search for alien life.........................

Most serious scientific searches for alien life look out past our Solar System and into deep space. There are billions of stars in the Universe. Many scientists figure that the chances of life somewhere else in the Universe are small but possible – and possible may be just good enough.

The Search for Extraterrestrial Intelligence (known as SETI) is an organization dedicated to finding life elsewhere. Among their many projects, SETI use radio telescopes and computers. What for? To monitor lots of radio frequencies for any possible radio signals sent to us. So far, no luck, but people remain hopeful.

In 1974 SETI sent a radio signal to a cluster of stars 24,000 light years away. The signal was sent in the form of pulses which made up a picture. The picture contained a number of things about Earth such as its position in the Solar System and what atomic elements make up life on our planet.

The message sent may seem crude, but could you do better?

Move to Mars—
COMMUNICATE WITH ALIENS

Make your very own message to send to aliens on another planet.

WHAT YOU'LL NEED
* a pen
* a piece of notepaper
* your brain in top gear

WHAT TO DO
What would you include on a piece of note paper to tell a creature who doesn't speak English all about you and the planet that you come from?

Pioneer and Voyager

Several space probes have been sent beyond Pluto and into deep space. *Pioneer 10* was the first man-made object to leave the Solar System — a journey which took nine years. It carried a plaque with pictures of a man and woman, a diagram of our Solar System and other information. By the time the *Voyager* space probe was launched in 1977, videodiscs had been invented and one was carried on board. It contained photos of Earth from space and various peoples of our world. It also included lots of sounds — from an elephant's trumpeting to greetings in different languages and music — even the sound of a kiss!

Now, the chances of *Voyager 11*'s videodisc falling into the hands of an intelligent life form are very, very slim. It is even less likely that an alien could possibly know what the disc is, play it and understand any of it. After all, who do you know with a videodisc recorder?

It's easy to mock, but scientists continue to hope that contact will be made. It would certainly be the most famous event of a whole millennium.

But what about Mars? ....

From the tantalizing glimpse of Mars that scientists have so far had from space probes and telescopes, one thing does emerge. It is 99.9 per cent certain that no intelligent alien life exists on Mars today.

How come?

Because the *Viking* missions that successfully landed on Mars proved that Mars has:

- ◯ no liquid water
- ◯ almost no oxygen
- ◯ intense cold
- ◯ no vegetation

Viking 1 and 2

In 1976 two *Viking* spacecraft parachuted safely down on to the surface of Mars. They made continuous measurements of temperature, wind speed and the length of the Martian day. Robot arms fitted to each lander scooped up soil samples and analysed them inside the craft. Much of what we know about the surface of Mars comes from these two spacecraft, which worked 7400 kilometres apart. Both worked on until the 1980s, when they were switched off.

What is keeping the scientific pulse racing about Mars, though, is whether life of any kind existed there in the past. Many scientists believe that Mars had lots of water flowing on its surface billions of years ago. The trick now is to find proof of life buried in the Martian rock.

The Martian Meteorite Mystery..............

One rock, a meteorite, caused an incredible fuss a few years ago. It was found in Antarctica, but it had come all the way from Mars an estimated 13,000 years ago.

I travel to Earth to escape the cold of deep space and look where I end up!

Labelled ALH84001, for a while this chunk of rock was in the news more than presidents and prime ministers. The reason was that it appeared to point to ancient life on Mars. But not all the scientists agree.

There is a chance that the features inside the meteorite may have been formed by a lifeless process. It is even more likely that they were formed by Earth bacteria after it had landed in Antarctica. Fierce debate is still raging.

Many scientists now believe that we need a mission to Mars that can actually return to Earth – to bring back fresh rock and soil samples for scientists to analyse. NASA is working on just such a mission, to be launched in about five years' time.

By the time it gets back, you might just have finished your astronaut training.

CHAPTER FIVE

GETTING THERE

The Universe is an absolutely enormous place. Most of it is out of our reach at the moment. Let's have a look at a few examples. If you turned your family car into a star car and travelled at a constant 160 km/h, it would take you 221000 MILLION years to reach the centre of our galaxy – the Milky Way. It would take you 3154 years just to get to Pluto, the most distant planet in our Solar System. But getting to Mars would be a breeze if you packed a big enough picnic – just 39 years and you're there. And spacecraft these days can travel a bit faster than 160 km/h.

Sending people to investigate Mars would be a massive achievement and it won't happen without a lot of hard work. Many unmanned missions have failed in the past. And although we've learnt from them, and now have far more powerful computers, unforseen errors can still occur. Just ask those involved with NASA's Climate Orbiter.

Here we go again
Mars, September 1999

The Climate Orbiter was designed to be Mars's first weather satellite. All was going well until September 1999. The craft had reached Mars's orbit, but something went wrong and signals went dead. The spacecraft got too close to Mars, and burned up in the Martian atmosphere. One reason for the crash was that one whole team of engineers was using metric units, while another was using feet and inches!

One advantage of manned missions is the cargo they carry – people. The first Mars voyagers will be supreme technical astronauts, able to perform many different tasks, including repairs to bits of the spacecraft that jam, buckle or simply stop working.

But that human cargo is also one of the biggest headaches. Unlike robots and automatic experiments, humans need to be fed and watered, supplied with air, toilets and somewhere to sleep. This all takes lots of space and weight.

The problem is that it takes a long time to get to Mars. The shortest journey to Mars would probably take around six months. That means carrying at least six months' supplies of food, water, air and other necessities to keep a spacecraft and its crew going.

And all this cargo has to be blasted clear of Earth.

Escaping Earth......................................

It's no easy task to produce enough oomph to get clear of Earth.

You have to reach a speed known as escape velocity to tear the spacecraft free of Earth's gravity – and it's pretty fast. How fast? 40,000 km/h or so. Quick enough for you?

He's not going to make escape velocity, you know.

Lots of ideas for launching future Mars missions have been suggested. One involves launching the mission from the Moon – which has only a tiny fraction of the gravity of Earth. This idea got a boost in 1998 when the probe Lunar Prospector found tell-tale signs of ice containing water at the Moon's poles. Water is made up of hydrogen and oxygen, and hydrogen and oxygen are just what you need to make rocket fuel.

Another idea involves using the International Space Station, being built at the moment, as a sort of giant construction and launch site. The many parts of the Mars mission craft would be hauled up to the station, assembled there and then launched from up in space.

To be honest, it's far more likely that the first manned Mars missions will use tried and trusted multi-stage rockets instead. These fire stage-by-stage and fall away when their fuel is used up, lightening the load to be sent into space. Want to see one in action?

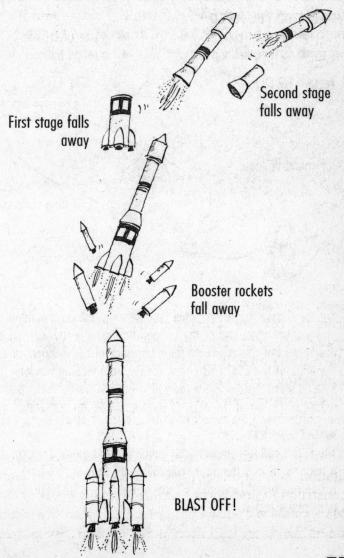

First stage falls away

Second stage falls away

Booster rockets fall away

BLAST OFF!

Move to Mars–
BUILD A MULTI-STAGE ROCKET

Here's how to make your very own multi-stage rocket, where parts fall away as they use up their fuel.

WHAT YOU'LL NEED
* a paper cup
* a long balloon
* a pair of scissors
* a round balloon

WHAT TO DO

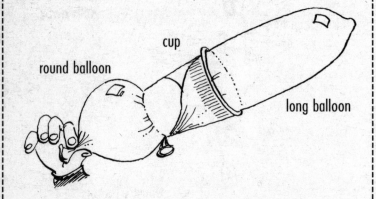

cup

round balloon

long balloon

Cut the bottom from the paper cup. Blow up the long balloon and pull the open end of the balloon through the top and out of the bottom of the cup. Fold the open neck of this balloon over the edge of the cup. This keeps the air from leaving the long balloon immediately. Place the round balloon inside the cup and inflate it. Let go of the neck of the round balloon and watch.

WHAT HAPPENS?
The round balloon propels your balloon rocket away. As soon as it deflates, it and the paper cup fall away, leaving the long balloon to speed forward. This set of balloons is just like a multi-stage rocket.

Weighty matters ....

The exact design of the rockets can be left to rocket scientists. What is important is the size and weight of what will be sent into space – experts call it the payload.

You can forget your hardback manuals and encyclopedias, especially when ten heavy volumes can fit with ease on a CD-ROM.

To reduce weight, a manned mission is likely to involve two, three or even more spacecraft. The craft with the crew on board will be launched on the quickest path available. Before then, robots will play a big part. They'll be sent ahead, travelling second class on slower cargo spacecraft to Mars. On arrival, the robots may set up machinery for making rocket fuel and other pieces of kit. They'll certainly check all the cargo is fine and radio back to base.

Several cargo missions, if successful, would provide all the supplies and equipment that the human crew members would need for their time on Mars.

Launch windows ....

And another important thing – you have to pick the right time to go.

How far Mars is away from us varies as both Earth and Mars make their trips on different paths around the Sun. There are key times when Mars is at its closest and other conditions are right. These are the best times to start a space mission to Mars, and they are called launch windows. Every 25 months or so, a reasonable launch window occurs. But the very best launch window – when Mars is both closest to the Earth and the Sun – comes around every 15 to 17 years. The next one of these is due in 2001.

Stay in touch ....

Once free of Earth, the manned craft would begin its five- or six-month trip on a carefully worked-out flight path to Mars. As it travels further and further away from Earth, it will take longer to send and receive messages. Once on Mars, the delay reaches ten minutes, as the successful 1997 Pathfinder/Sojourner mission to Mars proved.

Pathfinder and Sojourner

Mars, July 1997

Once it landed safely on Mars, the Pathfinder probe opened out like a flower. Its petals were solar panels. They had also been Sojourner's garage on the seven-month trip to Mars. Sojourner is the neat little rover that wandered around a tiny part of the Martian surface investigating rocks and soil.

➤

Sojourner sent information back to Pathfinder, and Pathfinder forwarded it to Earth, by means of radio waves.

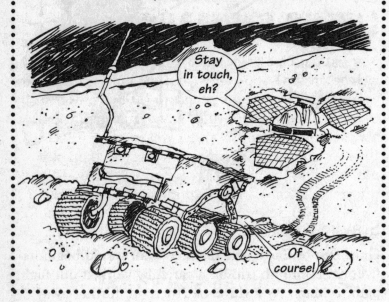

Landing party...

On reaching Mars, the craft would go into orbit around the planet before timing its descent and landing. Cushioned by parachutes, rocket blasters and possibly, even large balloons, the craft would land at its target area. A successful landing will trigger off wild celebrations back at mission control.

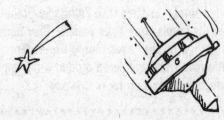

CHAPTER SIX

LIVING ON MARS

Mars would be a great place to live if it wasn't for a lack of air, food and plants, and the intense cold! How would it be possible for people to get by on Mars, even for a short while? And, more importantly, how are you going to survive when you're living there? It will take a lot of ingenuity, but experts think it is possible with the technology we already have. Information from unmanned missions like Global Surveyor will be vital in planning how people can survive on the red planet.

Global Surveyor
In orbit round Mars, now

Launched in 1996, Mars Global Surveyor (MGS) was over-shadowed by the Pathfinder mission. This is a great shame as, despite the odd hiccup, MGS has produced masses of information about Mars — data which may be very important to all the early visitors, including you.

Packed with high-tech kit including radar, lasers and advanced cameras, it was the first spacecraft to really analyse Mars's weather, how the polar ice-caps grow and shrink, and lots of other things too. It will certainly keep plenty of scientists busy.

Don't tell me. More data from Mars Global Surveyor.

Build your own extension ....

You and the other astronauts in a short stay mission to Mars might be able to get by living in your lander craft. This would probably be fitted with a couple of small, pressurised extensions which would look a bit like high-tech caravan awnings.

The key to these extensions and to larger buildings on Mars, would be to keep them full of air under pressure. This would mean that astronauts could live inside them without cumbersome pressure suits. Astronauts would enter and exit through airlocks that prevent any air from escaping and also stop the inside becoming contaminated with Mars dust.

YOU ARE NOW IN AIRLOCK

SUIT-VAC

Suits you..

To go outside on Mars, you would have to wear an advanced sort of spacesuit specially built for the unique Martian conditions. Mars-wear will have to be a lot lighter than the sorts of spacesuits and backpacks worn by astronauts visiting the Moon. Mars's gravity may be less than Earth's, but it's double that of the Moon's.

Developing a suit that is lightweight yet can be packed with enough supplies for four hours or more in the Martian atmosphere is a top priority for engineers on Earth. Food, water, power, communications and, most of all, something suitable to breathe, all have to be crammed in. And the suit must allow the astronaut to move freely. It's a tall order, but essential to a Mars mission.

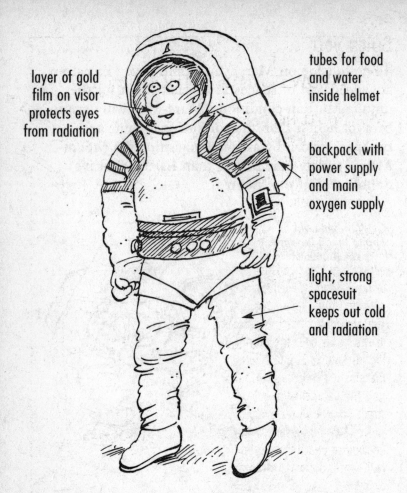

layer of gold film on visor protects eyes from radiation

tubes for food and water inside helmet

backpack with power supply and main oxygen supply

light, strong spacesuit keeps out cold and radiation

No soil toil ....

Perhaps the key experiment to be conducted on a Mars mission would be to try to grow plants. One technique is to do away with soil altogether. What do you grow the plants in then? Good question. The answer is a solution of water and nutrients. The technique is called hydroponics, and you can see how it's possible even without nutrients in the next experiment.

Move to Mars–
GROW PLANTS WITHOUT SOIL

WHAT YOU'LL NEED

* some green or brown lentils
* a glass jar
* an old pair of tights
* an elastic band

WHAT TO DO

Half fill the jar with water, add two or three teaspoons of lentils and leave them in a warm, dark place overnight. The next day, stretch a piece of the tights over the lid of the jar and fix in place with the elastic band. You can now turn the jar upside-down to strain out the water over a sink. Keep the jar in a warm, dark place and rinse the lentils with water a couple of times a day.

WHAT HAPPENS?

After four or five days, some of the lentils should start to sprout both shoots and roots. When seeds are soaked in water and left in the dark, an alarm bell rings inside the seed: 'Time to grow!' The food store inside the seed is enough to get the growing started. This gets used up though, and when you travel to Mars you will need to take nutrients with you to feed your plants as they grow.

You may also be growing your plants in inflatable greenhouses. The soil might be imported from Earth, or ground-up Mars rock with nutrients added. But where would the water come from? Well, it may be transported from Earth or generated by fuel cells, or you might get it by melting frost found under Mars's surface. Major investigations into the polar ice-caps will have to be made to see if a major water source exists under the surface.

Polar Lander
Mars, 1999

Polar Lander reached Mars in December 1999, but scientists are still working out why it never returned their calls. This is a great shame as it was sent to study the soil and look for ice beneath the surface of the Martian south pole. The mission featured two microprobes, called Deep Space 2, designed to smash into the soil and look for signs of frozen water ice. The microprobes hit the ground at a speed of 600 km/h. Ouch!

However water is first supplied, there's a good chance that it won't need that much topping up. Why? Because most of the water will be retained inside the Mars greenhouse. Want to see how? Try out the following simple experiment.

Move to Mars—
GROW PLANTS WITHOUT WATER

WHAT YOU'LL NEED
* a pot plant
* a clear plastic bag
* a twist tie

WHAT TO DO
Put the plastic bag over a leaf of the pot plant and gently tie it with the twist tie. Leave it for 24 hours and then have a look.

WHAT HAPPENS?
Inside the bag you will see water droplets. These have come from the leaf as it breathes out ('transpires' is the posh word for it). Water travels through a plant from the roots to the leaves. Most of the water that is breathed out by the leaves could be recycled and fed back to the roots.

Water isn't the only thing that would be recycled and reused on Mars. Nothing would be wasted at the Mars colony, simply because it costs too much to get it there in the first place. Even the contents of a full toilet may be treated with chemicals and then used as a sort of manure for the growing plants ...

Power play ....

Without power to drive the machines in a Mars colony, your mission's as good as over before it's started. Power is essential, and will probably be supplied in two or more different ways. If one power provider cannot work for some reason, another can kick in and compensate.

Where will the power come from? Well, there's a range of possibilities. You will probably bring a selection of fuel cells and batteries along from Earth. And before you even set off, experts will have told you whether it's going to be possible to make rocket fuel from the Martian atmosphere. If it is, then the landing craft will come with an electricity generator which runs off that type of fuel. Another possibility is a small nuclear reactor.

The most readily available source of energy would be solar power. Mars is a lot further away from the Sun, so the intensity of sunlight it receives is only half what we get on Earth. But – and it's an important but – that thin atmosphere of Mars will help. It lets a lot more of the Sun's rays through than Earth's heavy atmosphere. Solar energy is likely to be an important power provider for people on Mars, even though long dust storms may stop it working.

Getting around ....

Mars's rocky and unpredictable surface means that most travels will be on foot, close to the landing area. Mars colonists won't have a Ferrari parked at the base, but they probably will have a Mars rover.

The rover might have large, inflatable wheels, or tracks, and may have a special rocking body so that it can climb over sharp slopes without toppling over.

The other likely Mars rover won't carry people, but it will carry scientific experiments. An unmanned sailplane would be launched by small rockets and kept in the air, possibly by solar energy. For longer journeys, this glider-like craft could have its journey programmed in. But, closer to base, it could be remote-controlled by astronauts.

Staying long? ....

If your colony is a large one, and you plan to stay for a while, you will have to plan even more carefully. It will be impossible to transport everything a large colony needs from Earth. A different approach is called for.

The big idea in Mars studies at the moment is a philosophy called 'living off the land'. We know it would be impossible to live off Mars completely, but we may be able to nab enough important materials from its atmosphere and ground to make life a lot easier for ourselves.

In the future we may see oxygen and other useful gases being extracted from the Martian atmosphere, and possibly from water sources below the planet's surface.

There she blows! We've struck water!

In the longer term, Mars's surface contains a number of important materials which could be extracted and used. The Martian soil contains large amounts of gypsum, a material that can be used to make glass. It may be possible to crush the soil to make bricks, clays and cements. Silicon in the Martian soil could be turned into microprocessors – silicon chips!

CHAPTER SEVEN

TERRAFORMING

It's one thing to send small numbers of colonists to Mars to live in closed environments. It's quite another to make Mars a suitable place for millions of people to live. The technical term for shaping a planet to make it more like Earth is 'terraforming'. Whether terraforming is really possible, no one knows, but, in theory, it seems to be. One day – perhaps not too far in the future – we'll be ready to try it. Results won't be seen overnight – terraforming would take thousands of years.

Most people agree that there are three key things needed to get terraforming going:

    ✫ heat up the planet
    ✫ create a far thicker atmosphere
    ✫ introduce basic life forms

Bit chilly, this morning!

Brrrr! Turn up the heat ....

Heating up the planet wouldn't just mean Martian colonists not having to wear thermal underwear. It's hoped that it would melt much of the surface water believed to exist on Mars. It would also help to thicken up the atmosphere. A number of clever ways of warming up Mars have been suggested.

That wasn't one of them!

The key idea is to trap more of the Sun's energy on Mars. There are several different ways this could be done. Giant solar reflectors, many kilometres in size, could be built and put into space. These giants, covered in a sort of advanced tin foil, would redirect lots of the Sun's rays on to the polar ice-cap, melting part of it and releasing gases into the atmosphere.

Another way would be to spray dark dust (possibly from Mars's moon, Phobos) on to the polar caps. Dark colours absorb more heat than light ones, so a thin layer of dust might be enough to melt the poles.

Move to Mars—
TURN UP THE HEAT

You can see how tin foil reflects and black materials absorb the Sun's energy with the following experiment.

WHAT YOU'LL NEED
* four jars
* a piece of black paper
* some tin foil
* a piece of white paper
* some elastic bands
* a sunny day
* a thermometer

black paper   white paper   tin foil

WHAT TO DO
Wrap white paper around one jar, black paper around another, and tin foil around a third. Hold the materials in place with elastic bands, and fill the jars two-thirds full with water. Fill a fourth, clear jar with water and place all four in the Sun. After an hour, record the water temperature in all four jars.

Creating the right atmosphere ....

Thickening Mars's atmosphere will be absolutely
essential to terraforming success. Heating up Mars
will help release some gases into the atmosphere but
much, much more needs to be done. Many scientists

think that a problem that's occurring on Earth might offer a solution for heating up Mars. Global warming on Earth is believed to be the result of the greenhouse effect. This is where carbon dioxide and water vapour in the atmosphere act like a blanket, trapping much of the Sun's energy. This causes average temperatures to rise.

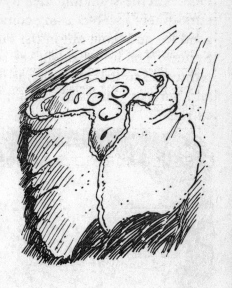

Earth's greenhouse effect is the result of hundreds of years of burning fossil fuels and cutting down large areas of forest. On Mars, it might be possible to engineer a similar effect by pumping large amounts of gases into the atmosphere. It would take a long, long time, but the gases may thicken the atmosphere, making it able to trap more heat and warm the planet a little more. What's more, a thicker atmosphere would mean more atmospheric pressure – something that most plants need to grow.

Sounds great, but where are these gases going to come from? Well, they can't all be transported from Earth. Perhaps processes will be invented which allow the gases to be made using materials from Mars's surface. Certainly melting parts of the polar ice-caps, which contain carbon dioxide, will help.

One extreme-sounding idea for adding gases to the atmosphere involves using comets, which are made of solid ice and hang out in the outer parts of the Solar System. The future might see several of these redirected so that they plunge into Mars's atmosphere and burn up – instantly adding gases. Told you it was extreme!

Hope for Japan

Mars, 2003

Mars's atmosphere, the little that there is of it, will shortly get a thorough checking from the first Japanese mission to Mars. The *Nozomi* (which means 'Hope') spacecraft is intended to hang around the red planet for a whole Mars year, to measure and map the atmosphere, assess Mars's weak magnetic field and beam back lots of photos of dust storms and the polar ice-caps.

➤

Engine problems which prevented the craft arriving in 1999 have been fixed. *Nozomi* should reach Mars in 2003.

Introducing basic life forms ....

No, not introducing basic life forms to each other, introducing them on Mars. Perhaps basic is unkind: these plants and microscopic creatures are among the hardiest of all the things that live on Earth. That's why they'd come first – but only after Mars has warmed up a little and starts to build its atmosphere. Earth soil is a good source of bacteria and microbes, as the next experiment shows.

Move to Mars–
SEE SOIL BREATHE

WHAT YOU'LL NEED
* some garden soil
* a small dish of limewater (ask at school or at a chemist's)
* a plastic box with a lid

WHAT TO DO
Place a few tablespoons of soil in the box and then pop a small dish or pot of limewater inside, next to the soil. Put the box lid on and leave.

WHAT HAPPENS?
Limewater is a liquid which goes cloudy if there is carbon dioxide gas around. After a couple of days, the limewater will turn cloudy, showing that there is carbon dioxide in the box. Where does it come from? From the thousands of tiny creatures in the soil which breathe out carbon dioxide.

Some of these microbes and certain plants can survive in low temperatures with little moisture and low atmospheric pressure. Scientists know – they've been experimenting. The bacteria and hardy plants which make the first trips to Mars will be their relatives. But, they will be genetically modified to be even more comfortable in Martian conditions and to give out as much oxygen as possible. Dark-coloured plants will be a speciality because as they flourish and cover parts of Mars they will absorb more rays from the Sun and will help heat the planet further.

Over time – and we mean thousands of years – larger plant and animal species may be introduced to Mars. The planet may even spawn its own unique plant and animal life. We won't see it, but future generations might …

CHAPTER EIGHT

LET'S GET COLONIZING

You've handed in the last of your homework. You've cancelled the milk and your comics. You're ready to travel with the other members of the first Mars colony. How will it all happen?

Step 1: On target...................................

Well, first of all, there will be lots of missions by the sorts of probes and landers you've already read about. These will map the surface of Mars in incredible detail, which will help scientists to decide the site of the first colony. More missions, equipped with advanced robots, will perform loads of experiments and try out prototypes of equipment. Testing this gear is so, so important. Prototypes of rocket-fuel makers and plant-growing kits must work and work well for a manned Mars colony to be feasible.

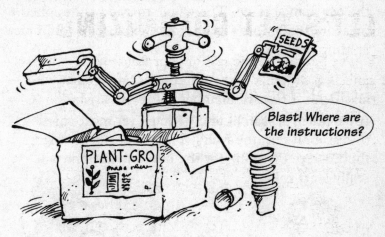

Mars Surveyor 2001

The concept of making rocket fuel on Mars may be tested out very soon. One of the experiment packages on board the *Mars Surveyor*, which will land on Mars in 2002, will see if it is possible to synthesize rocket fuel from the gases in the Martian atmosphere.

Mars Surveyor is a step on from the *Pathfinder/Sojourner* mission. It too will feature a robot crawler, deployed by a robot arm, which will explore part of the Martian surface. The whole mission is designed to measure the hazards that human explorers would face landing on Mars.

Step 2: The go-ahead...........................

After much research, and perhaps yet more missions to test it all out, the Mars colony will get the big thumbs up. Planning for the human missions will really start rolling.

Hooray!

First of all, large fleets of unmanned spacecraft carrying robots and essential kit will head off to Mars. Landing at the target site, the robots will construct and test the gear needed for the first Mars colony. Some will be mobile rovers, which will scan every square centimetre of the target site, mapping the features in incredible detail.

The robots will also assemble buildings and equipment. Supervising robots will relay pictures and other information back to mission control on Earth.

Step 3: Train your brain ....

Back on Earth, the excitement will be building. As one of the first potential colonists, you will be an instant celebrity.

Not that you'll have much time for talk shows, newspaper interviews or TV quizzes …

You'll be far too busy training for all that sort of stuff. You'll need to be physically fit, understand all the procedures of the mission and know your own responsibilities inside-out. Every member of the first colony teams will be mission specialists with their own important jobs. You may be the in-flight navigator, the crew's doctor, a space engineer or a geologist. You'll also have to learn the key parts of another crew member's job. This is so that if one person falls sick, there's always cover.

Step 4 Lift-off! ....

The launch window has arrived. Finally, the green
light will come. You'll be allowed to say goodbye to
your nearest and dearest, then, it's into the craft,
buckle your seat-belt and take a deep breath. You're
off to Mars!

Now, did I
pack my
toothbrush?

Step 5: In flight ....

Lift-off is the most exciting ride of your life. You feel
the forces thrusting you back into your seat as the
giant rockets propel you out of Earth's orbit. After the
euphoria of a successful lift-off, the next few hours
prickle with tension. Is everything working OK? Is the
mission on course? Has mission control detected a
problem? Is the craft going to be called back to Earth?

Once you know all is well and the ship's bang on
course, life settles down. You are on a journey that
will make that four-hour trip in the school minibus
seem like a picnic. Six months at least stands between
you and Mars. Hope you packed a few good books on
CD-ROM.

In fact, you've got plenty to do. You're keeping fit by working out in space, you have assignments to perform on the ship, and you're keeping yourself in training for your vital tasks on Mars …

… but not necessarily all at the same time.

Mission Control is like an annoying parent. They want to know absolutely everything about you day in, day out. This includes physical stuff, like your blood pressure, but also how you're feeling.

For the hundredth time, I told you, I feel fine.

Step 6: Happy landings ....

Gradually, the excitement on board the ship builds as you get closer and closer to your destination. Entering Mars's orbit and landing on the surface is the most dangerous part of the mission. Powerful computers handle most of the descent stage, but the crew has an override button in case something goes wrong. If all goes well, a combination of parachutes and brake thrusters should land you softly and safely on Mars.

Step 7: Those first days on Mars ....

You won't be heading out on to Mars immediately. There'll probably be a day or so of checks and double-checks before you slip into your special outdoors suit. Which of you is going to be first down the ladder and on to the surface will have been decided months ago.

Whether you're first, second or eleventh, you'll never forget your first walk on the surface. But even that first walk will involve jobs such as checking the outside of your landing craft. Over the next few days, you and the rest of the crew will venture further afield, perhaps in your very own Martian rover.

Finding the equipment set up by the unmanned missions, and checking it all, will be top priority. Then you and the rest of the crew will start assembling more experiments and kit, such as inflatable greenhouses in which to grow plants.

Step 8: Settling down ....

You are likely to have several months on Mars. Then again, if you are very lucky – if all the equipment works, if water has been found, energy is easy to create and plants can be grown – you might be in for a longer stay. There's even the chance that you'll be joined by more Mars colonists.

By the time more craft arrive, you could be living in large buildings connected to the outside by airlocks. Every trip outside means several hours of getting into your spacesuit, getting used to the pressure in the airlock, and cleaning yourself and the airlock down afterwards.

But it won't be all hard work. However much mission control back on Earth want to get value for money, they know they can't work the first colonists into the ground. Leisure time will be carefully designed to keep you fit and happy.

Step 9: Terraforming ....

Terraforming may start after several centuries of Mars colonies. Experiments and fieldwork would discover the best way to boost the atmosphere, release water supplies and encourage natural plant life. Work on a scale never attempted before would gradually transform the red planet into a blue and green one. Work on terraforming may get a boost from advances in nanotechnology. This is the science of making machines to an incredibly tiny scale. How small?

Well, they're measured in atoms rather than centimetres. Millions of mass-produced nano-machines would make a willing workforce capable of transforming Mars into a lush world.

Step 10: The new world ....

Many thousands of years after you arrived, Mars may be unrecognizable as the cold, hard, rocky world you knew. The terraforming process may have worked its magic – turning Mars into a planet with flowing water, a thick atmosphere and lush plant life. Dozens of generations of Mars-dwellers will have come and gone and the unique Martian diet, low gravity, and other factors may produce differences in the way people from Mars and Earth evolve. Who knows, by that time, the human race may be in cahoots with alien species.

The thousands of Mars settlers may know all about you and the rest of the first colonists to arrive. You and your crew mates from that first ever Mars mission may have been elevated to the status of gods ...

Then again, your name may be forgotten, just like the very first caveman or woman who looked upwards and spotted a strange red star twinkling in the night sky.